essentials

essentials liefern aktuelles Wissen in konzentrierter Form. Die Essenz dessen, worauf es als „State-of-the-Art" in der gegenwärtigen Fachdiskussion oder in der Praxis ankommt. *essentials* informieren schnell, unkompliziert und verständlich

- als Einführung in ein aktuelles Thema aus Ihrem Fachgebiet
- als Einstieg in ein für Sie noch unbekanntes Themenfeld
- als Einblick, um zum Thema mitreden zu können

Die Bücher in elektronischer und gedruckter Form bringen das Fachwissen von Springerautor*innen kompakt zur Darstellung. Sie sind besonders für die Nutzung als eBook auf Tablet-PCs, eBook-Readern und Smartphones geeignet. *essentials* sind Wissensbausteine aus den Wirtschafts-, Sozial- und Geisteswissenschaften, aus Technik und Naturwissenschaften sowie aus Medizin, Psychologie und Gesundheitsberufen. Von renommierten Autor*innen aller Springer-Verlagsmarken.

Torsten Schmiermund

The discovery of the periodic table of the chemical elements

A short journey from the beginnings until today

Springer

Torsten Schmiermund
Frankfurt am Main, Germany

ISSN 2197-6708 ISSN 2197-6716 (electronic)
essentials
ISBN 978-3-658-36447-2 ISBN 978-3-658-36448-9 (eBook)
https://doi.org/10.1007/978-3-658-36448-9

This book is a translation of the original German edition „Die Entdeckung des Periodensystems der chemischen Elemente" by Schmiermund, Torsten, published by Springer Fachmedien Wiesbaden GmbH in 2019. The translation was done with the help of artificial intelligence (machine translation by the service DeepL.com). A subsequent human revision was done primarily in terms of content, so that the book will read stylistically differently from a conventional translation. Springer Nature works continuously to further the development of tools for the production of books and on the related technologies to support the authors.

This Springer imprint is published by the registered company Springer Fachmedien Wiesbaden GmbH, part of Springer Nature.
The registered company address is: Abraham-Lincoln-Str. 46, 65189 Wiesbaden, Germany

What You Can Find in This *essential*

- A historical outline of the development of the periodic table of the elements (PTE)
- Information on preliminary work and precursors of the PTE
- An explanation of the structure of the PTE
- Different variants of the representation

Contents

1 Introduction

One hundred and fifty years ago, in 1869, the "periodic system of the elements" was discovered. It was not developed by natural scientists—it was discovered just as the chemical elements were discovered. The periodic table is a "nature-given order" and was only put into a form by man.

For many, the periodic table is a symbol of the chemistry lessons at school, which are often perceived as difficult or even annoying. At the same time, however, it is also a symbol for "chemistry"—just as the red "**A**" (apothecary) or the 'green cross' is a symbol for pharmacies. Take a look for yourself at what is on offer with a periodic table printed on it: Mugs, mouse pads, shower curtains, bath towels, cushion covers, carpets, lunch boxes and much more.

For most people who do not have a strong affinity for chemistry, the PSE is probably just a complicated table in which the chemical elements are listed according to their atomic number, but unfortunately in rows of different lengths (periods), without the meaning behind it being obvious. Only on closer inspection do the columns (groups) reveal relationships between the individual elements. If you are familiar with the electron structure of the elements, if you know the structure of the atomic shell, you will understand why the periods have different lengths. As is so often the case in the natural sciences, it is worth taking a second look.

The periodic table is timeless. The discovery of the noble gases at the end of the nineteenth century did not shake it, they could simply be added. Newer discoveries about the structure from the inside of atoms at the beginning of the twentieth century helped to better understand the structure of the system. The quantum mechanical model of the atom did not harm PSE, and the "new" elements created by man between 1960 and 2016 also fit easily into this scheme. Besides, the value of this "table" can hardly be overestimated from a theoretical, practical and pedagogical point of view.

T. Schmiermund, *The discovery of the periodic table of the chemical elements*, essentials, https://doi.org/10.1007/978-3-658-36448-9_1

What You Should Know

If you are an "old hand", already well versed in chemistry: Go for it. Have fun with the book.

If you are more of a "chemical beginner", then you should at least be familiar with the atomic symbols ("chemical abbreviations") for the elements.

Furthermore, it is advantageous to have some prior knowledge on the subject of atomic structure. The brief explanations in this book can only serve as a repetition/refresher. If necessary, please refer to the literature listed at the end.

Reference

If you would like to take a closer look at the topic, please take a look at the bibliography. Here you will find, in addition to the literature used for this *essential*, further book tips.

2 History of the Periodic Table

As is usually the case in the natural sciences, the "discovery" of the periodic table of the elements (PTE for short) was neither straightforward nor the result of a spontaneous intuition. Let us first take a look at the preliminary work that was done and how PTE came about in the form we know today.

2.1 The Atomic Theory

Already Democritus (460–370 B.C.) explained around 400 B.C. that all substances are made up of "atoms" (Greek: *atomos* = indivisible). He extended the view of his teacher Leukipp of Milet (approx. 450 B.C.) by assuming that the atoms of different substances differ in shape, size and mass. Furthermore, these atoms could assume a certain position in relation to each other, collide with each other, unite and separate again. In this way Democritus was able to explain the variety of substances surrounding us. Decades after Democritus, the Greek philosopher Epicurus (341–approx. 270 B.C.) developed the atomic theory further and wrote ten theorems about the basic building blocks of the world:

- Nothing arises from what is not.
- Nothing dissolves into that which is not.
- The whole thing is infinite.
- The whole thing has always been the way it is now, and always will be.
- The whole is made up of the bodies and the void.
- There are two types of bodies: atoms and aggregates (atomic compositions).

T. Schmiermund, *The discovery of the periodic table of the chemical elements*, essentials, https://doi.org/10.1007/978-3-658-36448-9_2

- The atoms move without stopping.
- Atoms have only three things in common with sensual things: shape, volume and weight.
- The atoms are infinite in number, the void is infinite in extent.
- The atoms of identical form are infinite in number, but their forms are indeterminate in number, but not infinite.

Thereafter, the atomic theory fell into oblivion until it was renewed and improved by the French physicist Gassendi (1592–1655). The English chemist Robert Boyle (1627–1691) took this up again in 1661 in his seminal book "*The Sceptical Chymist*" and developed a further improved theory of atoms.

In his book, Boyle rejects the "doctrine of the four elements" and the "doctrine of the three principles" based on the writings of Paracelcus' that defined the ideas of chemistry/alchemy for several centuries.

Boyle introduced a new concept of elements that excluded fire, earth, water, and air, among others, as elements. He attributed the chemical changes to changes in the structure of differently shaped particles ("corpuscles") and assumed that the different substances are formed by the coming together of differently shaped corpuscles into different forms.

Only with the model of the English chemist John Dalton (1766–1844), however, can we speak of a modern, scientific view of the atom concept. Based on his investigations of gases and their solubility in liquids, he developed the first "scientific" atomic model.

The central implications of Dalton's atomic model, published in 1805, are:

- Substances consist of smallest, not further decomposable particles, the atoms.
- The atoms of different elements have different masses and have different properties.
- The atoms of an element are equal to each other in mass and chemical properties.
- Atoms of different elements can combine with each other, in the ratio of simple integers, to form compounds.
- When a compound decomposes, the atoms that remain unchanged can form the same or other compounds again.
- The atoms can be conceived as massive, matter-filled spheres.

Dalton laid the foundation for the laws of "constant and multiple proportions" with his investigations on the relative masses of atoms (with the reference value hydrogen = 1, since he identified its atoms as those with the lowest mass) and a table of the relative atomic masses of 18 elements and compounds published in 1805. With his composite "atomic compounds", he already anticipated the concept of molecules, which was not introduced until 1860 (On the further development of atomic theory, Chap. 3).

2.2 The Atomic Masses

The atomic masses (formerly: atomic weights) are not absolute masses in grams or kilograms—the numerical values would be much too small and unwieldy. They are comparative values that are linked to the absolute mass (in grams) via the SI unit of the amount of substance (mol) with the factor of Avogadro's constant ($N_A = 6.022 \times 10^{23}$ mol^{-1}). Atomic and molecular masses are therefore given in "units", which refer to an arbitrarily fixed base. A nice side effect of the link with the constant N_A is that one can replace the "units"—without having to change the numerical value—by the unit of measurement g/mol and thus obtain the relative atomic mass (A_r).

Definition of Amount of Substance

By the way, as of May 2019, the SI unit of mass is defined as:

> The mole, unit symbol mol, is the SI unit of the amount of substance. One mole contains exactly 6.022 140 76 $\times$ 10^{23} individual particles. This number corresponds to the fixed numerical value applicable to the Avogadro constant N_A, expressed in the unit mol^{-1}, and is known as the Avogadro number.
>
> The amount of matter, symbol n, of a system is a measure of a number of specified individual particles. A single particle can be an atom, a molecule, an ion, an electron, another particle or a group of such particles with a precisely specified composition.
>
> The former link with the carbon isotope ^{12}C is thus invalid.

The first table with atomic masses was published by J. Dalton in 1805. Dalton chose the lightest element, hydrogen, as the reference value with the value "1".

In 1814, the Swedish chemist J. J. Berzelius (1779–1848) published a table of the atomic masses of elements and compounds. In this table, he arbitrarily set the value of oxygen to 100, since he considered this element to be the "pivot of chemistry".

The German physicist J. Meinecke (1781–1823) and the English physician W. Prout (1785–1850) noticed that the atomic masses of many elements were obviously integer multiples of the atomic mass of hydrogen. This was taken as a reason to regard hydrogen as the original substance and to assign it the mass "1".

Here, however, the hydrogen molecule (H_2) was set equal to 1. This resulted, for example, in C = 6 for carbon, O = 8 for oxygen and S = 16 for sulphur.

It was not until 1858 that the Italian chemist S. Cannizzaro (1826–1910) succeeded in clarifying the difference between the atomic mass and the molecular mass, and he recognized that hydrogen exists as a diatomic molecule. This gave the hydrogen atom (H) the value "1" and the hydrogen molecule (gaseous hydrogen, H_2) the value "2".

At a large congress of chemists in Karlsruhe in 1860, the focus was on the clarification of terms such as "atom", "molecule", "equivalent", "atomicity" or "basicity", the draft of a uniform formula notation and criteria for the determination of atomic and molecular masses. Among others, Beilstein, Bunsen, Cannizzaro, Dumas, Kekulé, Stas and Wurtz took part—in addition to D. I. Mendeleev, L. Meyer and W. Odling. In the aftermath of the congress, atomic and molecular theory became more widely accepted and atomic mass determinations became increasingly accurate. Undoubtedly, these are reasons why the PTE was discovered only in the following years.

The Belgian chemist J. S. Stas (1813–1891) had become known for his precise atomic mass determinations in 1837–1840. His proposal to use (natural) oxygen as the reference element with mass 16.000 was adopted internationally in 1865. Even beyond the discovery of the three isotopes of oxygen (1929), oxygen remained the reference element in chemistry until 1961. The rule was: 1/16 ^{nat}O = 1 amu (***atomic mass unit***). From 1929 to 1961, physicists used a reference scale based on the ^{16}O isotope (1/16 ^{16}O = 1 amu). For about 30 years, chemists and physicists thus had slightly different atomic masses: The difference ^{nat}O to ^{16}O is 0.03 ‰.

Since 1961, the carbon isotope ^{12}C has served as the basis and 1/12 of ^{12}C = 1 u (***unified atomic mass unit***), for chemistry and physics alike. This results in atomic masses e.g.: ^{nat}H: 1.00797, ^{nat}O: 15.9994; ^{nat}C: 12.01115.

In the periodic systems illustrated in this book, some of which have very different structures, only the numerical values of the atomic masses—without a unit—are usually given. Please note that in some cases different reference values have been used.

2.3 Early Attempts at Systematization

Already in ancient times there were attempts to divide the substances surrounding us into groups. A distinction was made between metals, stones (hard), earths (soft) and volatile substances. In the middle of the eighteenth century, the need for a classification increased as more and more (inorganic) substances—and thus also elements—were discovered. Thus, between 1750 and 1790 11 elements (H, N, Cl, O, Mn, Ni, Mo, Te, W, U, Zr) were discovered, and from 1790 to 1817 another 17 elements (Na, K, Ba, Sr, Ca, Mg, Be, Cr, Nb, Ta, Pd, Rh, Os, Ir, I, Li, Cd). A sufficient number of elements for the establishment of an ordering system was thus available.

After the concept of elements had been introduced in 1789 by A. L. de Lavoisier (1743–1794), 2 years later J. B. Richter (1762–1807) had formulated the "Law of Equivalent Proportions", in 1803 Dalton had published his atomic theory and Avogadro his molecular theory and thus, among other things, the possibility arose to determine relative atomic masses, the theoretical and metrological basis was also available to set up such an element system.

2.3.1 Prout's Original Substance Theory

Already at the beginning of the nineteenth century (1815), the English physician William Prout (1785–1850) tried to establish a connection between the elements known at that time. He assumed that all elements originated from a so-called **primordial substance**. He considered hydrogen to be this primordial substance—not entirely without reason, as we know today.

If Prout's hypothesis were completely correct, then all atomic masses would have to be integer multiples of the atomic mass of hydrogen. If we set the atomic mass of hydrogen equal to 1, then according to Prout the atomic masses of all elements should be integers. This fits fairly well in many cases (carbon 12, nitrogen 14, oxygen 16, sulfur 32), but does not fit at all in others (lithium 6.9; chlorine 36.5; magnesium 24.3; krypton 83.7). In particular, the atomic mass determinations carried out with the greatest accuracy by J. J. Berzelius (1779–1848) and J. S. Stas (1813–1891) showed that there is no question of integerity for most elements.

Today we know that the nucleus of the hydrogen atom, the proton, is in fact the building block of all other elements, and that all elements are formed by fusion of the nuclei of lighter elements in stars—starting with the fusion of hydrogen nuclei into helium nuclei. Prout's hypothesis can therefore be seen as a kind of "diffuse precursor" of modern atomic theoretical ideas.

2.3.2 Döbereiner's Triad Theory

J. W. Döbereiner (1780–1849) considered the elements from a chemical point of view in his **triadic theory** published in 1829. He had noticed that the three halogens chlorine, bromine and iodine have the rounded atomic masses 35.5, 80 and 127. The sum of the masses of chlorine and iodine divided by two gives 81.25—and thus relatively exactly the atomic mass of bromine.

Similarly, the chemical and physical properties of bromine are intermediate between those of chlorine and iodine. Chlorine[1] is a yellow-green gas, bromine is a brown liquid and iodine is a dark purple—almost black—solid. The rounded densities are: chlorine (liquefied): 1.51 g/cm^3, bromine: 3.12 g/cm^3, iodine: 4.93 g/cm^3. The acid strength increases from HCl (pK_S −7) to HBr (pK_S −9) to HI (pK_S −11). In terms of reactivity, bromine lies between chlorine and iodine, as shown, for example, by the formation of the hydrogen compounds. Thus, per mol of HX are released: HCl 92.4 kJ, HBr 51.9 kJ, HI 4.7 kJ.

Döbereiner called a group of three chemically similar elements, whose atomic masses are related as shown, a triad. Other triads are, for example, the alkali metals lithium—sodium—potassium, the alkaline earth metals calcium—strontium—barium or the chalcogens sulphur—selenium—tellurium.

Based on the triadic theory, L. Gmelin (1788–1853) published a precursor of the periodic table as early as 1843 (Fig. 2.1).

2.3.3 De Chancourtois' Cylinder

The periodicity of the elements was first discovered in 1862 by the French geologist A.-E. Béguyer de Chancourtois (1820–1886). He placed a helical line at a 45° angle on the mantle of a cylinder and divided the circumference into 16 sections. Each section corresponded to one atomic mass unit. The interesting thing was that in this arrangement related elements often came to lie on top of each other. The periodicity was evident in the fact that after advancing one turn, similar elements were found on the same line. This arrangement is also known as **telluric helix** (lat. *tellus* = earth and greek *helix* = curved, "helical"). A section of this helix in a modernized form can be seen in Fig. 2.2.

[1] Fluorine was first produced in pure form in 1886.

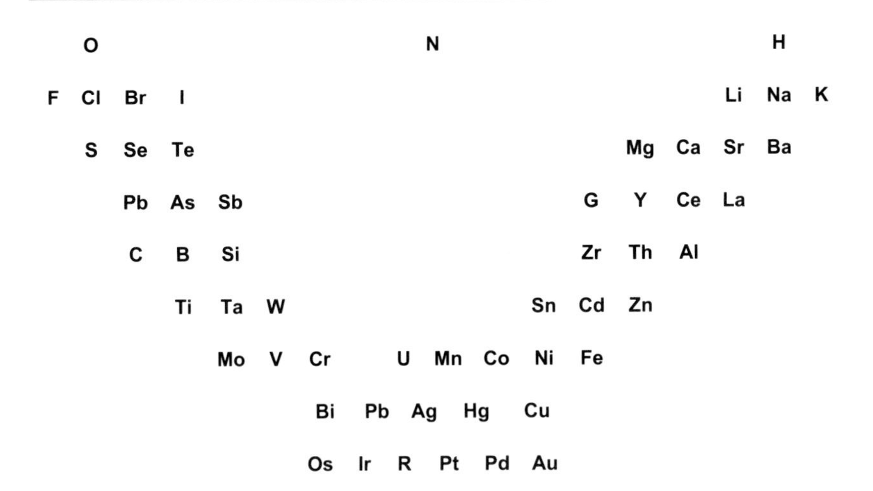

Fig. 2.1 Representation of the chemical elements. [After L. Gmelin (1843), new drawing]

2.3.4 Odling's Classification System

Even more rarely than Chancourtois (Sect. 2.3.3) and Newlands (Sect. 2.3.5), the English chemist W. Odling (1829–1921) is considered by historians as one of the pioneers of PTE.

As early as 1857, he divided the elements into 13 "natural groups" and created a series of element tables—precursors of the periodic table—from 1857 to 1865. In contrast to Newlands, he succeeded in classifying 57 of the 60 elements known at the time. Newland's system, on the other hand, included only 24 elements. Both Newlands and Odling published their lists in 1864 and independently came up with periodically repeating properties in 1865. Even though Odling had already developed a table of elements (and thus a precursor of PTE) in 1857, he did not publish his ideas until 1864, 2 years after Chancourtois. The similarity of Odling's design with the table published by Mendeleev in 1869 is striking (compare Fig. 6.1).

2.3.5 Newland's Octave Law

The work of J. A. Newlands (1837–1898), who rediscovered periodicity in 1864 (publication at about the same time as Odling), also led to the modern PTE. He ordered the elements according to increasing atomic mass and numbered them

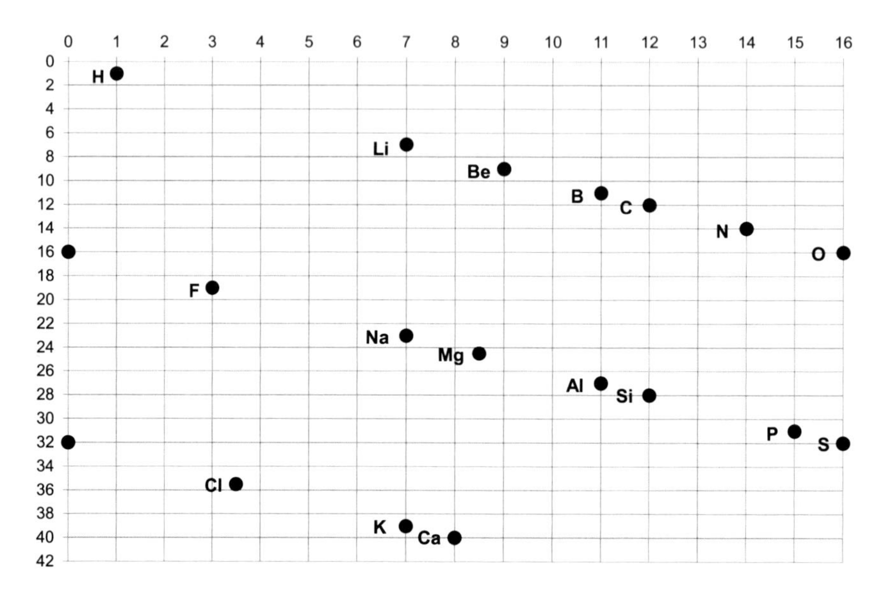

Fig. 2.2 Section of the "telluric helix". (After de Chancourtois 1862; redrawing)

consecutively (hydrogen = 1, lithium = 2, beryllium = 3, etc.—the noble gases were still unknown).

If he now started from any element, he arrived after seven elements at an eighth, similar element. This is similar to a musical scale, in which after every seven notes one comes to an eighth, similar note: the octave. Newlands therefore called the rule he found the "**law of octaves**". Since Newlands gave hydrogen a special position, there were first two periods with seven elements each. Then, however, from potassium to bromine, 14 already known elements had to be added. In the long run, this rule of eight proved to be too rigid and was discarded.

Newland's numbering, on the other hand, is and remains an integral part of the periodic table: the atomic numbers he introduced (albeit partly corrected) are still present.

2.4 Mendeleev's Table

The foundations of today's PTE were created by L. Meyer and D. I. Mendeleev independently of each other. Meyer created a divided table of 52 elements in 1864–1868 for his textbook "*Die modernen Theorien der Chemie* ("*The Modern Theories of Chemistry*"")", focusing on the physical properties (melting point, boiling point, density, etc.). Mendeleev, on the other hand, considered mainly the chemical properties.

Mendeleev was awarded the chair of technical chemistry at St. Petersburg University in 1865, focusing on lectures in inorganic chemistry. When he wrote his textbook "*Grundlagen der Chemie*" ("*Fundamentals of Chemistry*") in 1868, he felt compelled to arrange the chemical elements in some scientifically justifiable form. Mendeleev wrote the elements and their atomic masses on individual cards and regrouped them again and again until he finally found the right arrangement.

His basic idea was to regard the property "mass" as unchangeable in order to be able to relate all other properties to it. It was therefore clear to him to look for a dependence between the properties of the elements and their (atomic) mass.

If the elements are now arranged according to increasing atomic mass, a periodic repetition of their properties results. After a certain number of steps, one comes across an element which has similar properties to a previous one. After the same number of steps, one again encounters a similar element. This recurrence of similar things is called periodicity. It is after this that this principle of order got its name.

In his treatise "Die Beziehungen zwischen den Eigenschaften der Elemente und ihren Atomgewichten" ("*The Relations between the Properties of the Elements and their Atomic Weights*"; Mendeleev 1869b) of March 1869, Mendeleev's periodic table was published for the first time. The content of the comprehensive treatise can be summarized as follows:

1. The elements arranged according to the size of their atomic masses show a distinct periodicity in their properties.
2. Elements that are similar in chemical behavior have either similar atomic masses (e.g., Pt, Ir, Os) or uniformly increasing ones (e.g., K, Rb, Cs).
3. The arrangement of the elements in groups corresponds to their so-called valence.
4. The elements that occur frequently in nature have a low mass.
5. The size of the atomic mass determines the character of an element.

6. It is to be expected that some currently unknown elements will be discovered. For example, an element similar to aluminium or silicon with atomic masses of 65–75.
7. The atomic masses assumed so far can be subjected to a correction if corresponding analogous elements become known.
8. Some analogies of the elements can be found because of the size of the atomic mass.

Only 2 years later Mendeleev published a new table which had been extended by the elements discovered in the meantime (Mendeleev 1871).

In this newer table (Fig. 2.3), hydrogen is alone in the first row. In the second row follow: Lithium (monovalent, alkali metal, strongly basic), Beryllium (divalent, alkaline earth metal, strongly basic), Boron (trivalent, forms a weak acid), Carbon (tetravalent, forms a weak acid) and Nitrogen (pentavalent, forms a strong acid). Oxygen and fluorine are a little out of line. After these two nonmetallic elements, a striking change in properties follows in the third row. Sodium (monovalent, alkali metal, strongly basic) and magnesium (alkaline earth metal, basic) resemble the elements in the second row. The properties repeat. The irregularities in groups VI and VII also disappear: phosphorus (pentavalent, acid-forming) is followed by sulphur (hexavalent, acid-forming) and chlorine (septavalent, acid-forming).

In this way the system is divided into eight families or groups of elements. The lower groups contain the distinctly metallic elements, the higher groups the non-metals (not including Group VIII). Within the groups the metallic character increases from top to bottom. A line from boron to iodine separates the metals from the nonmetals.

If one compares the oxides, the valency towards oxygen increases according to the group number: Na_2O, MgO, Al_2O_3, SiO_2, P_2O_5, SO_3, Cl_2O_7. At the same time, the acidic character increases from left to right. Na_2O forms with water the strong base sodium hydroxide ($NaOH$), MgO the base magnesium hydroxide ($Mg(OH)_2$). SiO_2 forms the extremely weak silicic acid (H_2SiO_3). Diphosphorus pentoxide (P_2O_5) reacts to form the medium strong phosphoric acid (H_3PO_4) and SO_3 or Cl_2O_7 form the strong acids sulfuric acid (H_2SO_4) and perchloric acid ($HClO_4$). Al_2O_3 stands with its behavior in the middle between the bases and the acids: It governs amphoteric.

In the case of hydrogen compounds, the valency decreases regularly from group IV onwards: SiH_4, PH_3, H_2S, HCl. Here, too, the acidic character increases with increasing group number. SiH_4 is not an acid, PH_3 an extremely weak, H_2S a weak, HCl a strong acid.

Rows	Group I R_2O	Group II RO	Group III R_2O_3	Group IV RO_2 RH_4	Group V R_2O_5 RH_3	Group VI RO_3 RH_2	Group VII R_2O_7 RH	Group VIII RO_4
1	**H** (1)							
2	**Li** (7)	**Be** (9)	**B** (11)	**C** (12)	**N** (14)	**O** (16)	**F** (19)	
3	**Na** (23)	**Mg** (24)	**Al** (27)	**Si** (28)	**P** (31)	**S** (32)	**Cl** (35,5)	
4	**K** (39)	**Ca** (40)	–	**Ti** (48)	**V** (51)	**Cr** (52)	**Mn** (55)	**Fe** (56) | **Co** (59) | **Cu** (63,6)
5	(Cu)	**Zn** (65)	–	–	**As** (75)	**Se** (79)	**Br** (80)	
6	**Rb** (85)	**Sr** (88)	**Y** (89)	**Zr** (91)	**Nb** (94)	**Mo** (96)	–	**Ru** (102) | **Rh** (103) | **Pd** (107) | **Ag** (108)
7	(Ag)	**Cd** (112)	**In** (115)	**Sn** (119)	**Sb** (120)	**Te** (128)	**I** (127)	
8	**Cs** (133)	**Ba** (137)	**La** (139)	**Ce** (140)	? Di	–	–	– – – –
9	–	–	–	–	–	–	–	
10	–	–	**Yb** (172)	–	**Ta** (182)	**W** (184)	–	**Os** (191) | **Ir** (193) | **Pt** (195) | **Au** (197)
11	(Au)	**Hg** (201)	**Tl** (204)	**Pb** (207)	**Bi** (208)	–	–	
12	–	–	–	**Th** (232)	–	**U** (239)	–	– – – –

Fig. 2.3 Periodic table of Mendeleev (1871); redrawing

In addition to the results of the Karlsruhe Conference (cf. Sect. 2.2), Mendeleev benefited from the fact that 60 of the 92 naturally occurring elements had already been discovered. It may also have been very helpful that of the 17 naturally occurring rare earths (lanthanides and actinides) only five had been discovered: Ce, Yb, Di, Th, and U. His work would have been immeasurably more difficult because the lanthanides behave so similarly chemically that they are difficult to separate. Thus, the element Di or its oxide (Fig. 2.3, group V, line 8) later turned out to be a mixture of the oxides of Sm, Gd, Pr and Nd.

2.5 Mendeleev's Predictions

Mendeleev was so convinced of his system of order that he corrected atomic masses and made predictions about previously undiscovered elements. In his opinion, regularities did not tolerate exceptions.

Already when setting up his system he assumed that the atomic mass of beryllium was inaccurate or wrong. The atomic mass of Be equivalent to hydrogen was determined to be 4.5. That is, when molecular masses were determined, a value was found that was 4.5 times the equivalent value for hydrogen.

Because of its chemical similarity to aluminium, beryllium was initially assumed to be trivalent. The atomic mass was therefore $4.5 \times 3 = 13.5$. In Mendeleev's table, however, there is no place for metallic beryllium between the nonmetals carbon (atomic mass 12) and nitrogen (atomic mass 14). Mendeleev therefore claimed that beryllium must be divalent and have an atomic mass of 9 ($4.5 \times 2 = 9$). Accordingly, Be found its place between lithium and boron. It was not until 15 years later that this assumption proved to be correct. Similarly, he also corrected the atomic masses of In, U, Ti, Ce, Os, Ir and Pt.

Mendeleev's system left a number of "gaps". He claimed that there must be elements here that had not been discovered at that time. And not only that. He even made predictions about the properties of these elements and assigned tentative names. Below boron should be Eka-Boron,[2] below aluminium Eka-Aluminium and below silicon Eka-Silicon. During Mendeleev's lifetime, eka-aluminum (gallium) was discovered in 1875, eka-boron (scandium) in 1879, eka-silicon (germanium) in 1888, and eka-tellurium (polonium) in 1898. Figure 2.4 compares the predictions with the properties found in reality using the example of eka-silicon/germanium.

[2] *Eka* (Sanskrit) = 1.

	Prediction Eka-Silicon (Es)	Found Germanium (Ge)
Element		
Atomic mass	72 g/mol	72.3 g/mol
Density	5.5 g/cm³	5.409 g/cm³
molar volume	13 cm³/mol	13.2 cm³/mol
Appearance	dark grey	grey to silvery white
Melting point	infusible	sublimates at red heat without melting
Extraction	from the oxide by reduction	Reduction of the oxide with hydrogen
Oxide		
Formula	EsO_2	GeO_2
Density	4.7 g/cm³	4.703 g/cm³
Chloride		
Formula	$EsCl_4$	$GeCl_4$
State	liquid	liquid
Boiling point	90 °C	86 °C
Density	1.9 g/cm³	1.887 g/cm³
Sulfide		
Formula	EsS_2	GeS_2
soluble in	Ammonium sulfide solution	Ammonium sulfide solution

Fig. 2.4 Mendeleev's prediction using the example of eka-silicon/germanium

2.6 Mendeleev or Meyer?

After Meyer and Mendeleev had achieved great success with their periodic system, the French—for patriotic reasons—first propagated de Chancourtois, the British Newlands as the discoverer of periodic laws.

Very many scientists, on the other hand, regarded L. Meyer as the founder of the "periodic system". Already in 1864—in connection with the work on his textbook "*Die modernen Theorien der Chemie*"—he had arranged 28 elements in six columns, whereby the boron group and the—not yet discovered—noble gases were missing. He presented a further 22 elements, namely many metals, in separate tables, as he was initially unable to produce a single table for the 50 elements. Here Meyer still spoke of "relations" within the groups he established—**not** of periodicity. In 1868 he handed over a revised, handwritten version of the element table, which was probably intended for the second edition of his textbook (published in 1872), to his successor. In December 1869—i.e. 9 months after Mendleev—Meyer wrote a paper with an improved table (Meyer 1870) and the presentation of an "atomic volume curve" (Fig. 4.1). Here Meyer also spoke of periodic properties for the first time. However, unlike Mendeleev, he did not dare to predict the properties of as yet undiscovered elements and/or their compounds.

Although both attended the congress in Karlsruhe in 1860, it seems that they had not known each other personally. Meyer had also not read Mendeleev's work of

March 1869 before publishing his paper and only became aware of it in the course of 1870.

On the other hand, Mendeleev knew neither Meyer's textbook nor his essay from 1864, but was familiar with the work of Odling—and probably also with that of Newlands and Chancourtois. Mendeleev had already in 1869 pointed out a periodic dependence of the element properties on the atomic masses—thus being the first of the two to describe periodicity. He also changed—probably still in 1869—the table (published in 1871) to the effect that what is today called the second period ranged from sodium to fluorine—instead of from beryllium to sodium.

When Meyer realized that Mendeleev's system was broadly in agreement with his design, a dispute arose over the authorship of the discovery of the "periodic system of elements", which was openly fought out until about 1880 and was also discussed throughout the chemical community of the time. This dispute was not settled until 1882, when both scientists were awarded the Davy Medal, the highest distinction in chemistry at the time.

In sum, it remains to be said that both have extended and improved knowledge of the ordering system of the elements in a fundamental way, but also that other natural scientists such as de Chancourtois, Newlands and Odling have contributed their share.

Due to the more comprehensive first elaboration, the first mention of periodicity and the precise predictions, Mendeleev's achievement was finally rated as the highest. Therefore, he is considered the "father of the periodic table".

2.7 Further Development of PTE

As Mendeleev himself repeatedly pointed out, his arrangement was not perfect. In order to maintain the chemical similarities, measured by atomic mass, iodine and tellurium, but also cobalt and nickel had to change places.

In the years 1892–1898 the noble gases (He, Ne, Ar, Xe, Kr) were discovered, but as a "zeroth group" (they do not form compounds, i.e. the valence is "0") they could nevertheless be fitted well into Mendeleev's system—whereby argon and potassium were also given "swapped" places.

The classification of the elements following lanthanum—the lanthanides—proved to be more difficult. When the elements between Ce (140 u) and Yb (172 u) were discovered, it became apparent that they were so similar to each other and to lanthanum that they could hardly be separated from it. A "chemical relationship" with Mo, Ag, Cd, etc. could not be established. At first, they were removed from the periodic table as a "special group".

For the further development of PTE, the discovery of radioactivity proved fruitful. On the one hand, it made it easier to fill the gaps that still existed before uranium. On the other hand, it led to deeper insights into atomic structure.

It turned out that there are substances that differ in their atomic mass by a few units, but are chemically identical. It was recognised that these were the same elements and these identical elements, which only differed in their mass, were called isotopes (Greek: *isos* = equal and *topos* = place; "located in the same place").

When isotopes were also found in the elements found in nature, the "crooked" atomic masses could also be explained. Chlorine, for example, consists of two isotopes: about 75% chlorine of mass 35 and just under 25% chlorine of mass 37. The average value is $35 \times 0.75 + 37 \times 0.25 = 35.5$, which is exactly the atomic mass that was found. Most elements occur as mixtures of isotopes; they are called mixed elements. Only 19 elements occur as so-called pure elements—i.e. only as a single isotope—in nature.

Due to the appearance of isotopes, however, the atomic mass had to be abandoned as an ordering characteristic. It was replaced by the atomic number, i.e. the "serial number" of the respective element. This was first introduced by Newlands (see Sect. 2.3.5) as a pure classification criterion. In 1913, the physicist H. Moseley found a connection between the atomic number (not the atomic mass!) and the X-ray radiation emitted by an element. It also turned out that Moseley's atomic number measurements agreed well with Rutherford's determinations of nuclear charge numbers: "In the atom there is a fundamental quantity which increases in regular steps from one element to another. This quantity can only be the positive charge of the atomic nucleus," Moseley stated.

Thus the atomic number became the number of positive nuclear charges and thus from a simplifying numbering to a fundamental ordering principle.

The correlations found by Moseley also led to the discovery of elements 72 (Hafnium, Hf, 178.5 g/mol) and 75 (Rhenium, Re, 186.2 g/mol) on the basis of their X-ray spectra.

Modern nuclear physics has filled a gap in the PTE by producing Tc. It has also (as of 2019) expanded it by an additional 24 elements: from element 95 Am (Americium) to element 118 Og (Oganesson).

The Atomic Structure

3

According to the ideas of the English physicist Sir E. Rutherford (1871–1939) from 1911, a positive atomic nucleus is orbited by (negative) electrons. The nucleus carries almost the entire mass of the atom and has only about one ten-thousandth of the diameter of the entire atom. The size of an atom is determined by its shell, in which the electrons are located.

Two years later, the Danish physicist N. Bohr (1885–1962) came to the conclusion that the electrons of the atomic shell are grouped in shells around the atomic nucleus. This led to a theory of atomic structure which—extended and supplemented—is still valid today.

3.1 Structure of the Atomic Nucleus

As we know today, the atomic nucleus, which accounts for practically the entire mass of an atom, consists of protons (sign: p, charge +1, mass: 1.0073 u) and neutrons (sign: n, charge ±0, mass 1.0087 u). The atomic nucleus measures on average about 15 fm (fm = femtometer = 10^{-15} m) in diameter.

The number of protons (= atomic number) determines which element it is. The neutrons act as a kind of "glue" that holds the equally charged protons together and thus makes a stable atomic nucleus possible (for more information, see e.g. Schmiermund 2019, p. 329 f.) The sum of the masses of the protons and neutrons determines the atomic mass of the respective atom.

In shorthand notation, the number of nucleons (= the sum of protons and neutrons) is written in superscript in front of the element symbol (^{xx}M), the number of protons (= the atomic number) in subscript in front of the element symbol ($_{xx}M$).

T. Schmiermund, *The discovery of the periodic table of the chemical elements*, essentials, https://doi.org/10.1007/978-3-658-36448-9_3

Examples for the Notation of the Element Symbol with Atomic Number Only
Hydrogen: $_{1}H$; Sodium: $_{11}Na$; Phosphorus: $_{15}P$; Silver: $_{47}Ag$; Mercury: $_{80}Hg$

3.2 Structure of the Atomic Shell

The atom itself is much larger than the nucleus because of the electron shell surrounding the nucleus: the average diameter is about 150 pm (pm = picometer = 10^{-12} m). Thus, the atom has a diameter about 10,000 times larger than the nucleus.

This "shell" surrounding the atomic nucleus consists of electrons (sign: e^-, charge −1, mass: 1/1836 of the proton). They are not only the carrier of the electric current, but are also of utmost importance for the bonds that atoms form with each other.

3.2.1 Atomic Model According to Bohr

N. Bohr made assumptions about the structure of atoms in 1913 (Bohr's postulates) and thus arrived at his atomic model. The following applies to this atomic model:

- The atom consists of an atomic nucleus and an atomic shell.
- The atomic nucleus is positively charged, carries almost the entire mass of the atom and is located in its centre.
- The atomic nucleus has only about 1/10,000 of the diameter of the whole atom.
- The atomic shell determines the size of the atom.
- The atomic shell is negatively charged. In it are the electrons of the atom.
- The electrons move only on certain ("allowed") paths around the atomic nucleus.
- The further out the orbits are, the more energetic the electron is.
- When changing from one orbit to another, energy is only absorbed or released in the form of "energy packets" (energy quanta).

Bohr first named the resulting electron shells with capital letters (from inside to outside: K, L, M, N, O, P). From Bohr's electron shells, a maximum number of electrons per orbit is also obtained. For this purpose, the Bohr orbits are numbered from the inside to the outside. The innermost orbit is numbered 1, the next outermost orbit 2, and so on. The maximum number of electrons is then given by $2n^2$, with n = shell number. Figure 3.1 shows the shell structure using the sodium

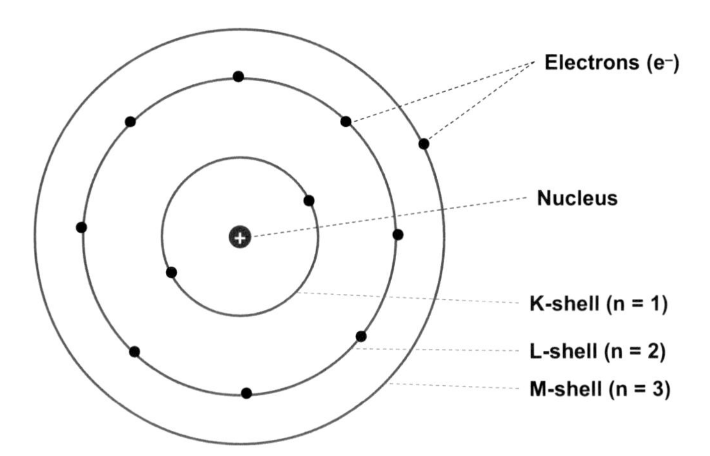

Fig. 3.1 Bohr atomic model for sodium

atom as an example. A representation of the PSE with Bohr's shells is shown in Fig. 3.2 for the first 14 elements.

3.2.2 Bohr-Sommerfeld Atomic Model

The Bohr atomic model marks the turning point to the quantum mechanical atomic model (also called the orbital model). Sommerfeld generalized the Bohr model and extended it. Among other things, four different so-called "quantum numbers" were introduced:

- Principal quantum number "n" is equal to Bohr's orbit and includes *nearly coincident* energy states of the electron.
- The subsidiary quantum number "ℓ" characterizes the energy states of the electron *within* the individual shells. Like the Bohr orbits ("main shells"), the subsidiary quantum numbers ("subshells" or "levels") are also designated by letters:
  $\ell = 0 \rightarrow$ s; $\ell = 1 \rightarrow$ p; $\ell = 2 \rightarrow$ d; $\ell = 3 \rightarrow$ f
- The magnetic quantum number "m" refines the splitting of the energy states by taking into account influences of (strong) magnetic fields (possible values: $-\ell$ ... 0 ... $+\ell$).

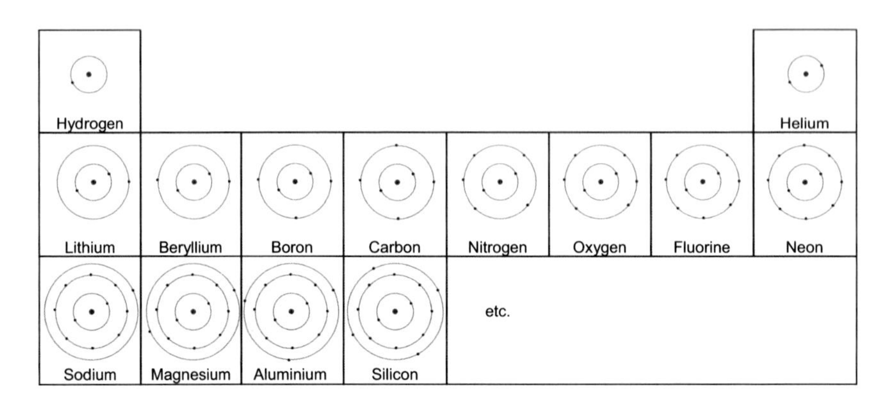

Fig. 3.2 Periodic table (detail) with representation of Bohr's atomic model

- The spin quantum number "*s*" indicates the "intrinsic angular momentum" of the electrons. The spin can only assume the values +½ and −½.

According to W. Pauli (1900–1958), two electrons in an atom must not coincide in all four quantum numbers. Considering this exclusion principle, one can now determine how many electrons the respective level (subshell) is capable of accepting:

Level	Possible secondary quantum numbers ℓ	Possible magnetic quantum numbers m	Possible spin quantum numbers s	Max. number of electrons
s	0	0	+½, −½	2
p	−1, 0, +1	−1, 0, +1	+½, −½	8
d	−2, −1, 0, +1, +2	−2, −1, 0, +1, +2	+½, −½	10
f	−3, −2, −1, 0, +1, +2, +3	−3, −2, −1, 0, +1, +2, +3	+½, −½	14

Now, in order to be able to specify the electron configuration of an element, the identification letter of the minor quantum number ℓ is preceded by the major quantum number n (the number of Bohr's orbit). The number of electrons in the respective level is written by a superscript digit:

Bohr's shell		Number of levels	Designation of the levels	Calculation max. number of electrons	Maximum electron distribution
K-shell	1. Shell	1	1s	$2 \times 1^2 = 2 \times 1 = 2$	$1s^2$
L-shell	2. Shell	2	2s, 2p	$2 \times 2^2 = 2 \times 4 = 8$	$2s^2\ 2p^6$
M-shell	3. Shell	3	3s, 3p, 3d	$2 \times 3^2 = 2 \times 9 = 18$	$3s^2\ 3p^6\ 3d^{10}$
N-shell	4. Shell	4	4s, 4p, 4d, 4f	$2 \times 4^2 = 2 \times 16 = 32$	$4s^2\ 4p^6\ 4d^{10}\ 4\ f^{14}$

Resulting electron configurations are often written—to shorten the overall sequence—in such a way that the preceding noble gas is placed in front of the configuration. Here are a few examples:

Element	Electron configuration (long form)	Electron configuration (short form)
Hydrogen (H)	$1s^1$	$1s^1$
Helium (He)	$1s^2$	$1s^2$
Beryllium (Be)	$1s^2\ 2s^2$	[He] $2s^2$
Oxygen (O)	$1s^2\ 2s^2\ 2p^4$	[He] $2s^2\ 2p^4$
Chlorine (Cl)	$1s^2\ 2s^2\ 2p^4\ 3s^2\ 3p^5$	[Ne] $3s^2\ 3p^5$
Potassium (K)	$1s^2\ 2s^2\ 2p^4\ 3s^2\ 3p^6\ 4s^1$	[Ar] $4s^1$
Krypton (Kr)	$1s^2\ 2s^2\ 2p^4\ 3s^2\ 3p^6\ 3d^{10}\ 4s^2\ 4p^6$	[Ar] $3d^{10}\ 4s^2\ 4p^6$

3.2.3 Orbital Model

A closer look reveals that electrons do not move in fixed orbits, but rather stay within certain regions around the atomic nucleus with extremely high probability. These areas are called orbitals. The principal quantum number n makes a statement about the size of the orbital, the minor quantum number ℓ about the shape, and the magnetic quantum number m about the spatial orientation of the orbital. A more detailed explanation can be found, for example, in Schmiermund (2019, p. 86 f.).

3.3 Occupation of the Atomic Shell with Electrons

In the periodic table, the elements are arranged from left to right according to increasing nuclear charge number. It is in the nature of things that with every proton added to the nucleus, an electron must also be added to the shell in order to obtain a neutral atom.

However, the 3d-orbital has a higher energy content than the 4p-orbital. Therefore, after the 4s orbital, the 3d orbital is filled first, followed by the higher-energy 4p orbital. In order to remember or derive this order of occupation, the checkerboard method is recommended (compare Fig. 3.3). If one assigns this occupation sequence to the elements and chooses the same arrangement as in the PSE, the result is Fig. 3.4.

This order applies *almost* without restriction. Deviations arise with half-occupied or fully occupied energy levels of the d- and f-levels, respectively, since these have a higher stability.

Examples of Deviations in the Cast Order

- $_{28}$Ni: [Ar] $3d^8$ $4s^2$ → **$_{29}$Cu: [Ar] $3d^{10}$ $4s^1$** → $_{30}$Zn: [Ar] $3d^{10}$ $4s^2$
- $_{63}$Eu: [Xe] $4f^7$ $6s^2$ → **$_{64}$Gd: [Xe] $4f^7$ $5d^1$ $6s^2$** → $_{65}$Tb: [Xe] $4f^9$ $6s^2$
- $_{78}$Pt: [Xe] $4f^{14}$ $5d^9$ $6s^1$ → **$_{79}$Au: [Xe] $4f^{14}$ $5d^{10}$ $4s^1$** → $_{80}$Hg: [Xe] $4f^{14}$ $5d^{10}$ $6s^2$

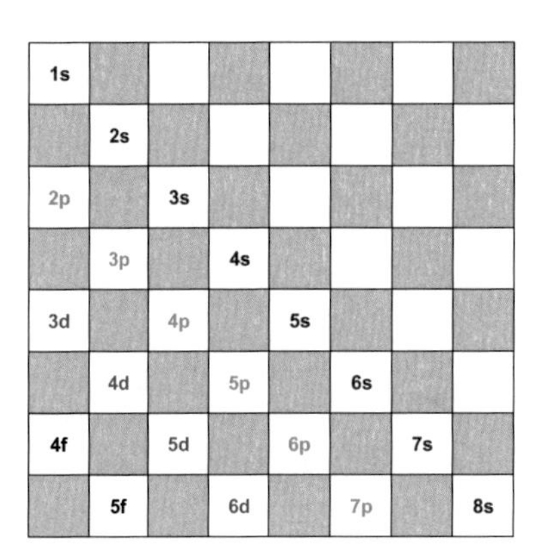

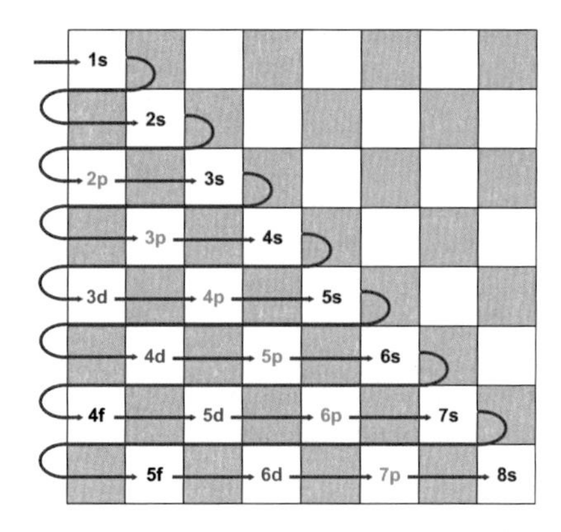

Fig. 3.3 Orbital occupation according to the "checkerboard" method

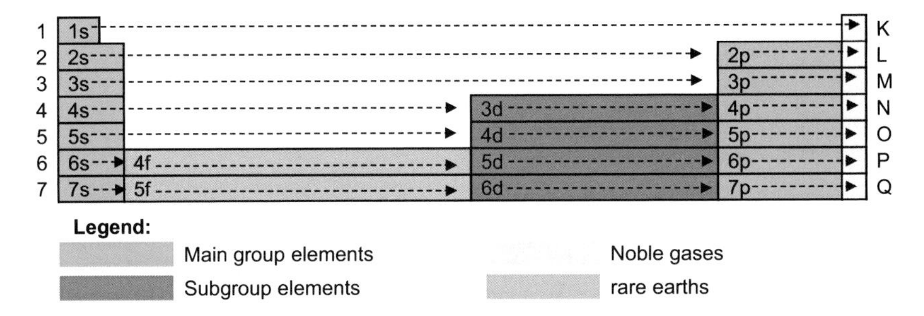

Fig. 3.4 Overview representation of the occupation sequence as PSE

4 The Periodicity

Both chemical and physical properties of the elements depend periodically on their atomic mass—more precisely: on their atomic number; even more precisely: on their electron structure. Some of these properties will be examined in more detail in order to better understand this periodicity. It is important to remember that the concept of periodicity is central to inorganic chemistry. The PSE systematizes chemical facts and thus helps to make them understandable.

The actual reason for the periodicity of various properties is a combination of nuclear charge, the number of electron shells and the number of outer electrons, i.e. the occupation of the individual levels with electrons.

4.1 Periodicity of Physical Properties

Periodic trends in the physical properties of the elements were not recognized until 1868 by L. Meyer's atomic volume curve and subsequently studied. Chemical properties and repetitive trends in the elements, on the other hand, were already the subject of scientific investigation from the end of the eighteenth century.

4.1.1 Atomic Volume

Figure 4.1 shows a modernised version of Meyer's atomic volume curve. One can clearly see the maxima for the alkali metals and the minima, which lie around the middle of the respective period (C, Al, Co, Ru, Ir).

T. Schmiermund, *The discovery of the periodic table of the chemical elements*, essentials, https://doi.org/10.1007/978-3-658-36448-9_4

In the alkali metals, a new outer shell occupied by an electron is added. In the middle of the period, the maximum number of bonding electrons is present, and the radius is smallest. With more electrons, the space requirement slowly increases again, since these additional electrons also take up space.

4.1.2 Ionisation Energy

Ionization energy is the energy that must be expended to remove an electron from the electron shell of an atom or ion. The low ionization energy of the alkali metals (Fig. 4.1) explains, for example, their high reactivity and flame colors. Furthermore, one can see:

- The elements of the first main group (Li, Na, K, Rb) have very low ionization energies. The single outer electron (electron configuration: s^1) is easy to separate.
- For the elements of the second main group, the energy is somewhat higher. Here, an electron must be removed from the fully occupied s orbital (electron configuration s^2).
- In the third main group (B, Al, Ga, In) the ionization energy is lower again. The single electron of the p-orbital (electron configuration s^2 p^1) is again easier to remove.
- Finally, the ionization energy of the noble gases is unusually high. The electron configuration s^2 p^6 (or s^2 for He) represents a very stable state.

4.1.3 Electron Affinity

Electron affinity is the energy converted by a neutral atom when it accepts an electron. Energy that is released by the absorption of an electron is represented by positive values in Fig. 4.2a. Energy that must be expended is shown as a negative value.

- The halogens (F, Cl, Br, I) take up electrons very easily. They then form the more stable noble gas configuration s^2 p^6 from their original electron configuration s^2 p^5.
- The slightly higher affinity of chlorine to fluorine is due to the fact that the increasing nuclear charge from fluorine to chlorine (from 9 p^+ to 17 p^+) exceeds the effect of the increasing atomic radius (from 64 pm to 99 pm).

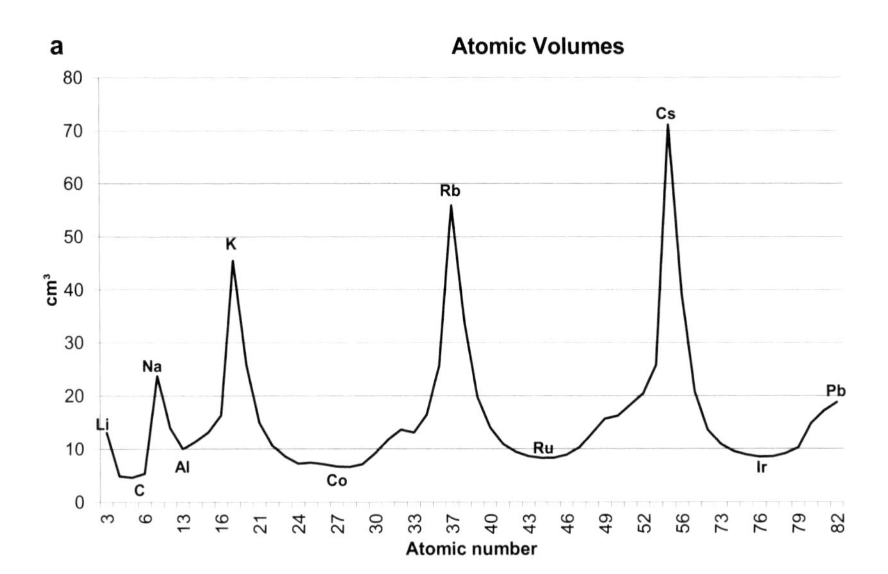

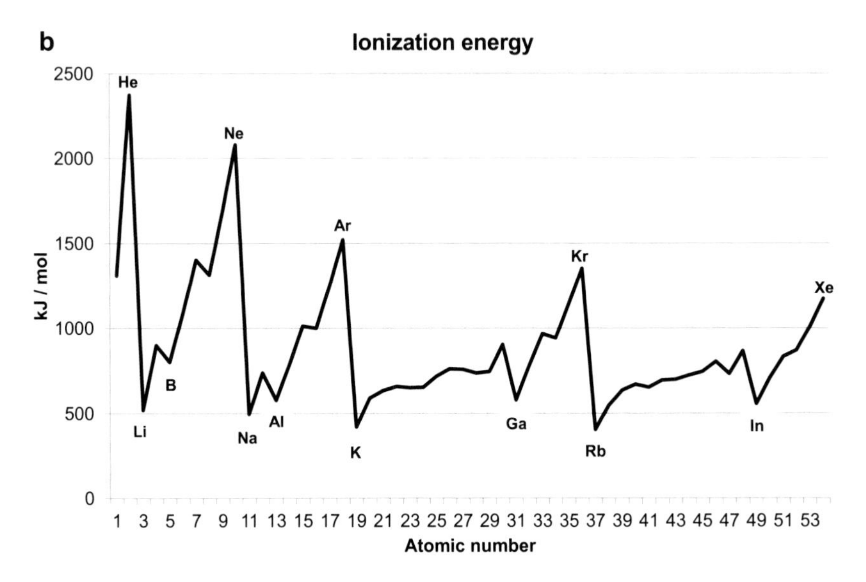

Fig. 4.1 Periodicity: (**a**) Atomic volume curve. (After L. Meyer; revised); (**b**) Ionization energy

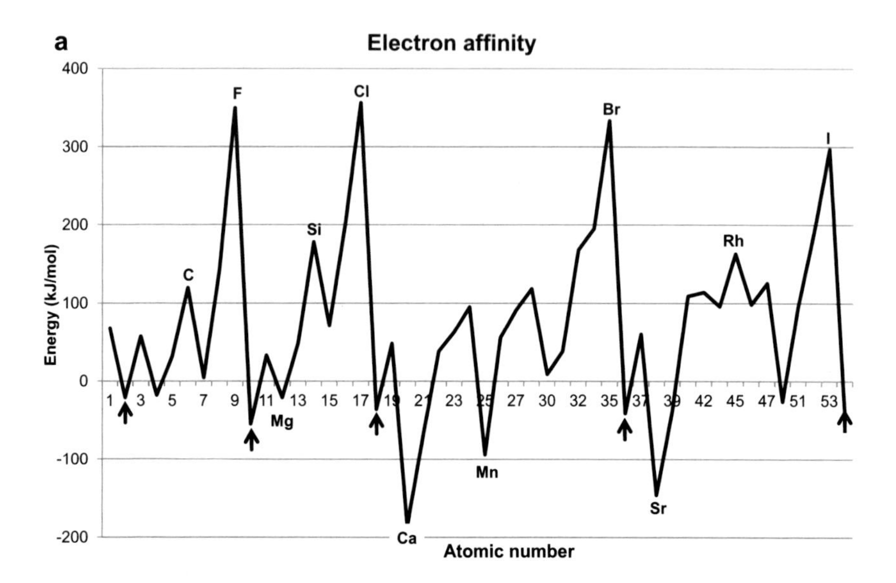

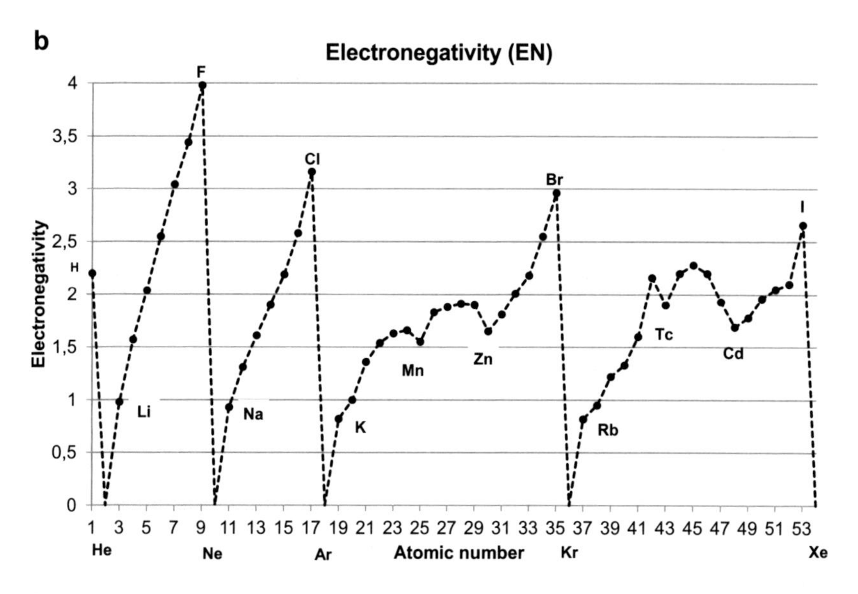

Fig. 4.2 Periodicity: (**a**) Electron affinity; (**b**) Electronegativity (EN)

- The alkaline earth metals (Be, Mg, Ca, Sr) can only with great difficulty add another electron to their atomic shell, since a p-orbital must be formed in addition to the fully occupied s-orbital.
- The energies of the noble gases (arrows) are all about the same, since the formation of a "new shell" (the next s-orbital is occupied by an electron) requires a comparably high energy input.

4.1.4 Other Periodic Physical Properties

It is left to the inclined reader to work out the periodicity of further physical properties.

For the elements, these can be, for example: melting points, boiling points, density, heat of fusion/evaporation or the dissociation energies.

The comparison of the density or the melting and boiling points of simple binary compounds such as hydrides, chlorides, oxides or sulfides also shows the analogous curves and thus the periodicity of the respective property.

4.2 Periodicity of Chemical Properties

The chemical properties cannot be "put into figures", i.e. described quantitatively, as easily as is possible with the physical properties. Nevertheless, trends can be identified. On the one hand, if one compares the elements within their group, but also if one looks at the course along the period.

4.2.1 Electronegativity

L. Pauling (1901–1994) developed a system in 1932 to estimate the polarity of a bond. For this purpose he introduced a measure, the electronegativity. Electronegativity (EN) is a measure of the tendency of an atom to attract electrons to itself in a bond.

By fixing the EN of fluorine to the value 4.0, a closed electronegativity scale could be given for all elements. The EN plays an important role in the bonds between the atoms and thus in the formation of ions and molecules.

Since noble gases do not form compounds in the "classical sense", no electronegativities can be determined for them. In Fig. 4.2b the values were arbitrarily set to "0".

4.2.2 Valence and Oxidation Number

The so-called valence is a characteristic element property. It is often defined as the number of hydrogen atoms that can bond with the element to form the hydrides, or the number of chlorine atoms that can bond with the element to form the chlorides. Sometimes it is also defined as twice the number of oxygen atoms that will bond to the respective atom in oxides.

The main group elements (groups Ia–VIIIa) form hydrides of the general formula MH_n, where n is equal to the group number N (= 1, 2, 3, 4) for groups Ia–IVa and obeys the calculation VIII—N (= 3, 2, 1, 0) for groups Va–VIIIa.

For the oxides, increasing maximum valences occur, which are directly related to the group numbers. The following table and Fig. 4.4 illustrate this once again.

Group	I	II	III	IV	V	VI	VII	VIII
Hydride	MH	MH_2	MH_3	MH_4	MH_3	MH_2	MH	–
Oxides	M_2O	MO	M_2O_3	MO_2	M_2O_5	MO_3	M_2O_7	MO_4

The valence is first of all the ratio described here. Taking into account the electronegativities, this results in the so-called oxidation number. The more electronegative element receives a negative sign, the less electronegative ("more electropositive") element a positive sign. By general agreement, these are written as Roman numerals and with a sign *above* the respective element if it is necessary to specify the oxidation number. In Fig. 4.3, the oxidation numbers of periods 1–4 are shown—and written in Arabic numerals for better readability.

4.2.3 Acidic or Basic Character of the Oxides and Hydrides

The alkali metals react directly with water to form hydroxides according to $2\ M + 2\ H_2O \rightarrow 2\ MOH + H_2$. Their oxides also react: $M_2O + H_2O \rightarrow 2\ MOH$. On the other hand, when oxides of nonmetals react, acids are formed: $SO_3 + H_2O \rightarrow H_2SO_4$. A tendency can also be derived from this.

Thus, the hydroxides MOH of the I main group are all strong bases whose basicity increases downwards. The hydroxides of the VII main group are acids,

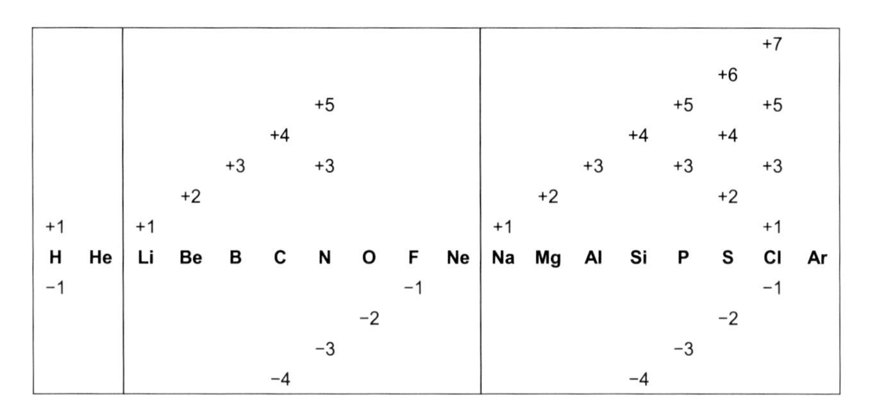

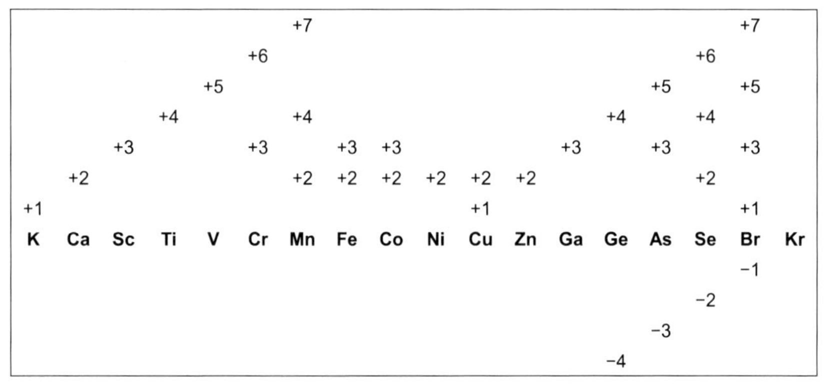

Fig. 4.3 Representation of the oxidation numbers of the elements of the first four periods

whose acidity decreases downwards. The elements in the middle of the main groups are—more or less pronounced—amphoteric (e. g.: $Be(OH)_2$, $Al(OH)_3$, $Ga(OH)_3$, Sn $(OH)_2$). The addition of oxygen atoms to the free electron pairs of the central atoms (which corresponds to an increase in the oxidation number) of groups V–VII noticeably increases their acidity, as shown here in the example of chloric acids:

Compound	HClO	$HClO_2$	$HClO_3$	$HClO_4$
Oxidation number Cl	+I	+III	+V	+VII
Hydroxide formula	Cl(OH)	ClO(OH)	$ClO_2(OH)$	$ClO_3(OH)$
Acid character	Weak	Medium	Strong	Very strong
pK_S value	7.54	1.97	−2.7	−10

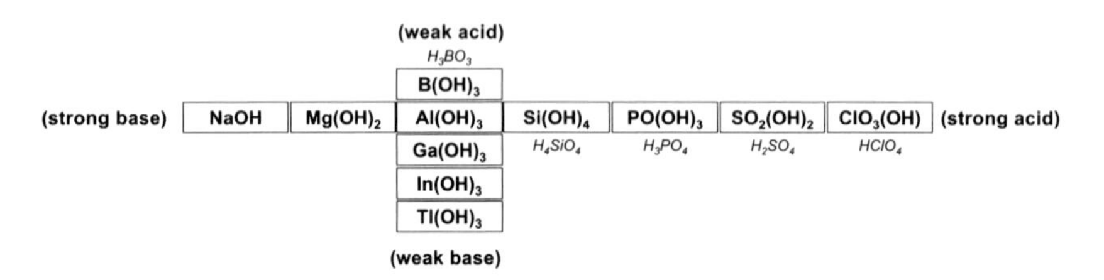

Fig. 4.4 Acidic or basic character of the hydroxides of the elements using the example of the third period

Figure 4.4 illustrates this again in general using the example of the third period.

It is different with the hydrides. The binary hydrides of the metals are bases, those of the nonmetals acids. The acid strength increases from the top to the bottom, as the comparison of the halogen hydrides clearly shows:

Compound	HF	HCl	HBr	HI
Acid character	Medium	Very strong	Very strong	Very strong
pK_S value	−3.14	−6	−8.9	−10

4.3 Summary of Periodicity

The most important periodic properties of the elements are listed in a clear table in Table 4.1.

Table 4.1 Important periodic properties of the elements

Atomic radius, atomic volume	•Increases from top to bottom •Decreases from left to right	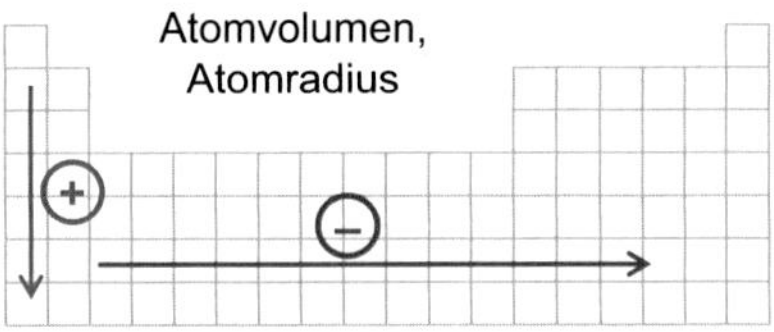 Atomvolumen, Atomradius
Ionization energy	•Decreases from top to bottom •Increases from left to right	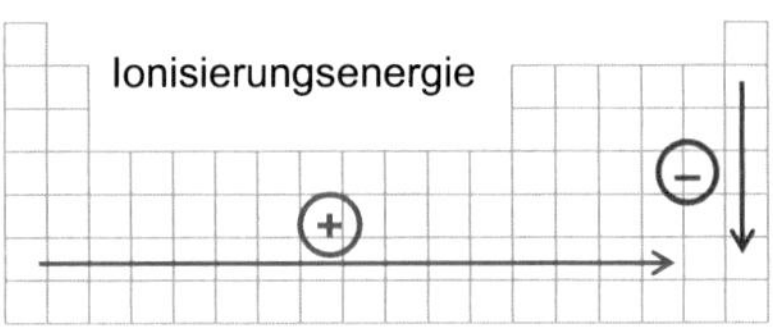 Ionisierungsenergie
Electron affinity	•Very high for halogens •Very low for alkaline earth metals	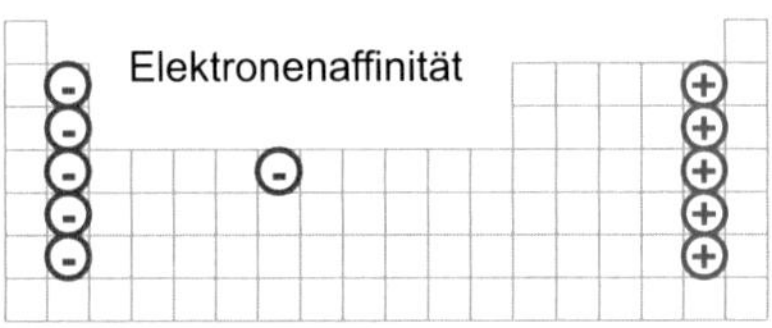 Elektronenaffinität
Electronegativity	•Decreases from top to bottom •Increases from left to right •Most electronegative element: fluorine ("most non-metallic non-metal") •Most electropositive element: francium ("most metallic metal")	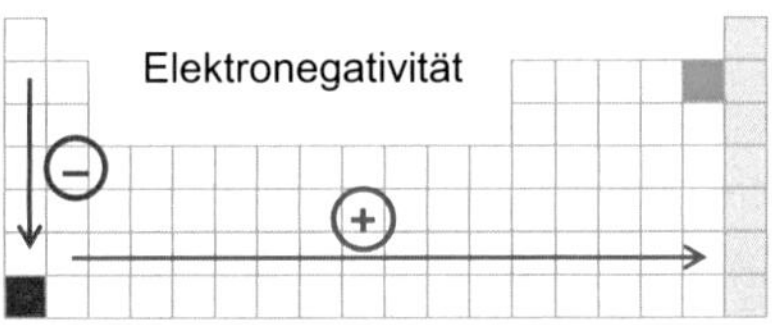 Elektronegativität
Metal character	•Increases from top to bottom •Decreases from left to right	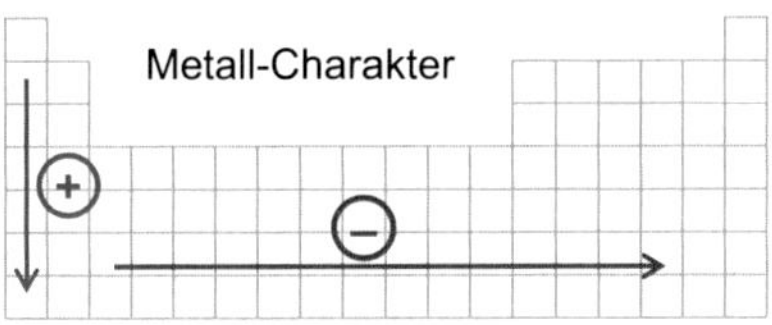 Metall-Charakter

5 Today's PSE

Today's periodic table is divided into seven periods of eight main groups and eight subgroups, whereby one subgroup (VIIIb) consists of three subgroups (8, 9, 10). This results in a total of 18 groups. In addition, there are two periods in which the so-called lanthanoids or actinoids—14 each—are added. However, these do not receive a separate numbering.

The horizontal rows of the periodic table, the periods, show an increasing number of order or electrons from left to right.

The vertical columns of the system are called groups. Due to the analogous configuration of the outer electrons, these elements show a chemically similar behavior and are therefore sometimes also called element families.

For many years, the CAS (*Chemical Abstracts Service*) numbering system was authoritative. Here, the main groups were given the suffix "a" and the subgroups the suffix "b". The Roman numerals corresponded—more or less—to the number of outer electrons.

For more than 25 years, the group numbering of the IUPAC (*International Union of Pure and Applied Chemistry*) has been bindingly valid. Here the groups are numbered from left to right from 1 to 18. Therefore, for the main groups 13–18, the number 10 has to be subtracted from the group number to determine the number of outer electrons. (Due to this "calculation", the old CAS numbering is still valid in some areas).

It is common to both numbering systems that the so-called rare earths (= lanthanides and actinides) do not receive group numbers. Table 5.1 gives an overview of the major and minor groups with their electron configuration and oxidation numbers.

For the currently valid representation of the periodic table (IUPAC, so-called "long form"), see Fig. 5.1. Note that, in contrast to earlier representations, the

T. Schmiermund, *The discovery of the periodic table of the chemical elements*, essentials, https://doi.org/10.1007/978-3-658-36448-9_5

Table 5.1 Overview of the main and subgroups of the periodic table

Group (CAS, old)	Group (IUPAC)	e-configuration (outer electrons)	Group name or elements of the group	Frequent oxidation numbers
Main groups				
Ia	1	s^1	Alkali metals Li, Na, K, Rb, Cs, Fr	+I
IIa	2	s^2	Alkaline earth metals Be, Mg, Ca, Sr, Ba, Ra	+II
IIIa	13	s^2p^1	Boron group B, Al, Ga, In, Tl, N(h)	+III
Iva	14	s^2p^2	Carbon group C, Si, Ge, Sn, Pb, F(l)	+IV/−IV
Va	15	s^2p^3	Nitrogen group N, P, As, Sb, Bi, (Mc)	+V/+III/−III
Via	16	s^2p^4	Chalcogenes O, S, Se, Te, Po, (Lv)	+VI, +IV, +II, −II
VIIa	17	s^2p^5	Halogens F, Cl, Br, I, At, (Ts)	+VII, +V, +III, +I, −I
VIIIa	18	s^2p^6	Noble gases He, Ne, Ar, Kr, Xe, Rn	0
Subgroups				
IIIb	3	s^2d^1	Sc, Y, La, Ac	+III
IVb	4	s^2d^2	Ti, Zr, Hf, (Rf)	+IV
Vb	5	s^2d^3	V, Nb, Ta, (Db)	+V
VIb	6	s^1d^5	Cr, Mo, W, (Sg)	+II, +III, +VI
VIIb	7	s^2d^5	Mn, Tc, Re, (Bh)	+II, +IV, +VII
VIIIb	8	s^2d^6	Fe, Ru, Os, (Hs)	+II, +III
	9	s^2d^7	Co, Rh, Ir, (Mt)	+II, +III
	10	s^2d^8	Ni, Pd, Pt, (Ds)	+II, +IV
Ib	11	s^1d^{10}	Cu, Ag, Au, (Rg)	+I (+II), (+III)
IIb	12	s^2d^{10}	Zn, Cd, Hg, (Cn)	+II

elements La and Ac form the first element of the lanthanide and actinide series, which is placed at the bottom. In older versions, lanthanum and actinium were often assigned to group 3 (old: IIIb) and the following elements were "inserted".

Ia	IIa	IIIb	IVb	Vb	VIb	VIIb	VIIIb			IIb	Ib	IIIa	IVa	Va	VIa	VIIa	VIIIa
1	2	3	4	5	6	7	8	9	10	11	12	13	14	15	16	17	18
H																	He
Li	Be											B	C	N	O	F	Ne
Na	Mg											Al	Si	P	S	Cl	Ar
K	Ca	Sc	Ti	V	Cr	Mn	Fe	Co	Ni	Cu	Zn	Ga	Ge	As	Se	Br	Kr
Rb	Sr	Y	Zr	Nb	Mo	Tc	Ru	Rh	Pd	Ag	Cd	In	Sn	Sb	Te	I	Xe
Cs	Ba	57-71 (La)	Hf	Ta	W	Re	Os	Ir	Pt	Au	Hg	Tl	Pb	Bi	Po	At	Rn
Fr	Ra	89-103 (Ac)	Rf	Db	Sg	Bh	Hs	Mt	Ds	Rg	Cn	Nh	Fl	Mc	Lv	Ts	Og

La	Ce	Pr	Nd	Pm	Sm	Eu	Gd	Tb	Dy	Ho	Er	Tm	Yb	Lu
Ac	Th	Pa	U	Np	Pu	Am	Cm	Bk	Cf	Es	Fm	Md	No	Lr

Fig. 5.1 Present-day periodic table, current IUPAC representation (so-called "long form")

6 Other Forms of the Periodic Table

The periodic table has also undergone many changes over the years. There is no *optimal* form of the periodic table. The choice of the form of representation is also determined by the purpose of the representation itself. These purposes can be, for example: chemical similarities, the electron configuration, the occupied shells and subshells, and so on.

Feel free to make the effort and do some research on the Internet for "periodic systems" and "alternative periodic systems"—or look at the URL at the end of the bibliography. You are sure to find interesting variations and ideas for handicrafts.

6.1 Periodic Tables in "Old Form"

In the first periodic tables, the "groups" were arranged row by row, the "periods" column by column. This is true of both Odling (1864) and Mendeleev (1869), as shown in Fig. 6.1.

T. Schmiermund, *The discovery of the periodic table of the chemical elements*, essentials, https://doi.org/10.1007/978-3-658-36448-9_6

a

			Ti = 50	Zr = 90	? = 180
			V = 51	Nb = 94	Ta = 182
			Cr = 52	Mo = 96	W = 186
			Mn = 55	Rh = 104,4	Pt = 197,4
			Fe = 56	Ru = 104,4	Ir = 198
			Ni = Co = 59	Pd = 106,6	Os = 199
H = 1			Cu = 63,4	Ag = 108	Hg = 200
	Be = 9,4	Mg = 24	Zn = 65,2	Cd = 112	
	B = 11	Al = 27,4	? = 68	Ur = 116	Au = 197?
	C = 12	Si = 28	? = 70	Sn = 118	
	N = 14	P = 31	As = 75	Sb = 122	Bi = 210?
	O = 16	S = 32	Se = 79,4	Te = 128?	
	F = 19	Cl = 35,5	Br = 80	J = 127	
Li = 7	Na = 23	K = 39	Rb = 85,4	Cs = 133	Tl = 204
		Ca = 40	Sr = 87,6	Ba = 137	Pb = 207
		? = 45	Ce = 92		
		?Er = 56	La = 94		
		?Yt = 60	Di = 95		
		?In = 75,6	Th = 118 ?		

b

			Ro 104	Pt 197
			Ru 104	Ir 197
			Pd 106,5	Os 199
H 1	"	"	Ag 108	Au 196,5
"	"	Zn 65	Cd 112	Hg 200
Li 7	"	"	"	Tl 203
Be 9	"	"	"	Pb 207
B 11	Al 27,5	"	U 120	"
C 12	Si 28	"	Sn 118	"
N 14	P 31	As 75	Sb 122	Bi 210
O 16	S 32	Se 79,5	Te 129	"
F 19	Cl 35,5	Br 80	I 127	"
Na 23	K 39	Rb 85	Cs 133	
Mg 24	Ca 40	Sr 87,5	Ba 137	"
	Ti 50	Zr 89,5	Ta 138	Th 231,5
	"	Ce 92	"	
	Cr 52,5	Mo 96	V 137	
	Mn 55		W 184	
	Fe 56			
	Co 59			
	Ni 59			
	Cu 63,5			

Fig. 6.1 Periodic tables in line-by-line representation (**a**) Mendeleev (1869), (**b**) Odling (1864)

6.2 Periodic Table According to Thomsen and Bohr

Thomsen (1895) and Bohr (1923) proposed a version of a pyramid rotated by 90°. Here the periods are arranged as vertical rows. Elements that come to stand below each other in today's PSE are connected with lines (Fig. 6.2).

The top line are the alkali metals, below them the alkaline earth metals. The bottom line are the noble gases, above them the halogens. The outgroup elements (d-elements) are framed in the third and fourth columns, respectively. The rare earths (lanthanides and actinides, f-elements) in the last two columns are double-framed.

6.3 Loops and Spirals

Kipp proposed a spiral arrangement in 1942. Elements of the same group are marked by the same colors. The upper, small loop contains the subgroup elements, the lower, larger loop the main group elements. Neither the transuranics nor the elements following lanthanum and actinium, respectively, are included in this representation, so uranium (proton number 92) is the last element in this representation. In Fig. 6.3a, the group numbers according to CAS are given in brackets in addition to the IUPAC group numbering.

Theodor Benfey designed a loop-period system (Fig. 6.3b) which is a further development of Kipp's design and includes all the elements known today.

6.4 Short-Period Systems

So-called short-period systems were characterized by the fact that main and subgroups were arranged in the same grid. This goes back to Mendeleev's second version (1871, cf. Fig. 2.3). Today, this form of representation is no longer in use. Figure 6.4 shows a typical short-period system around 1950. Major and minor groups are arranged in the same column and shifted to the left ("a") or right ("b"). The lanthanoids are written as an additional line below, the actinoids not yet discovered.

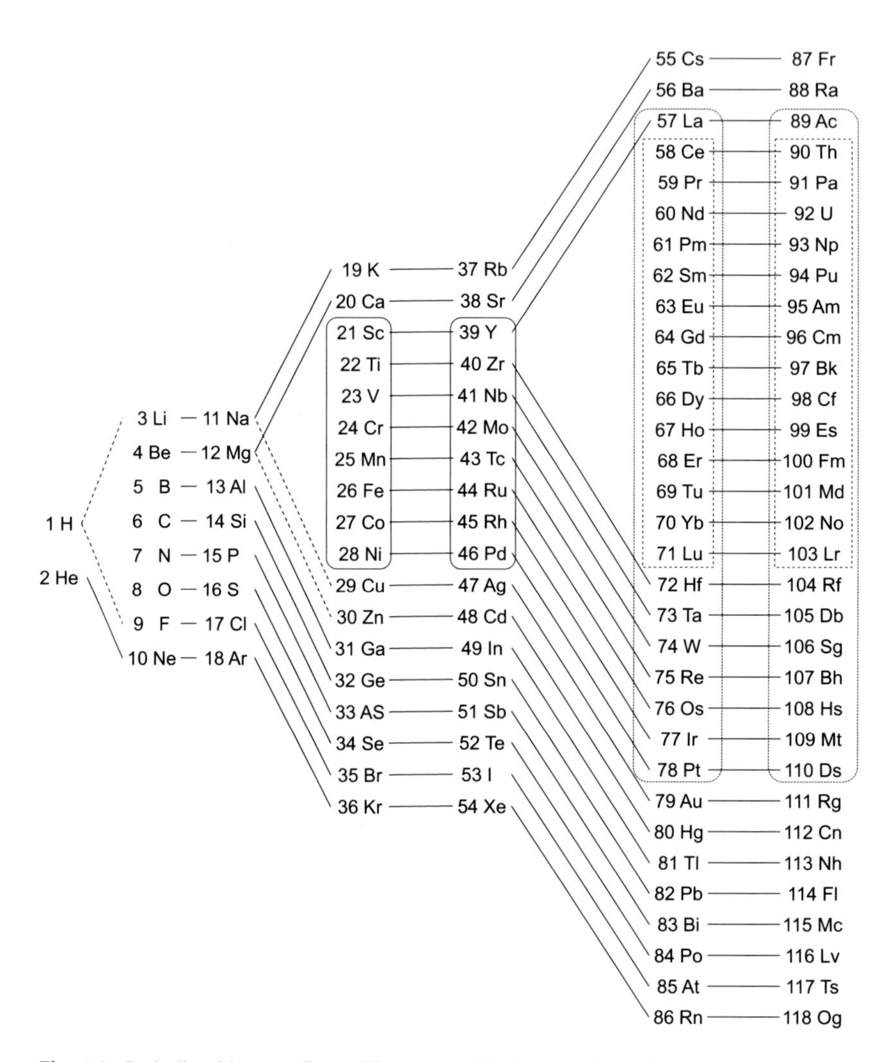

Fig. 6.2 Periodic table according to Thomsen and Bohr, extended redrawing. (For details see text Sect. 6.2)

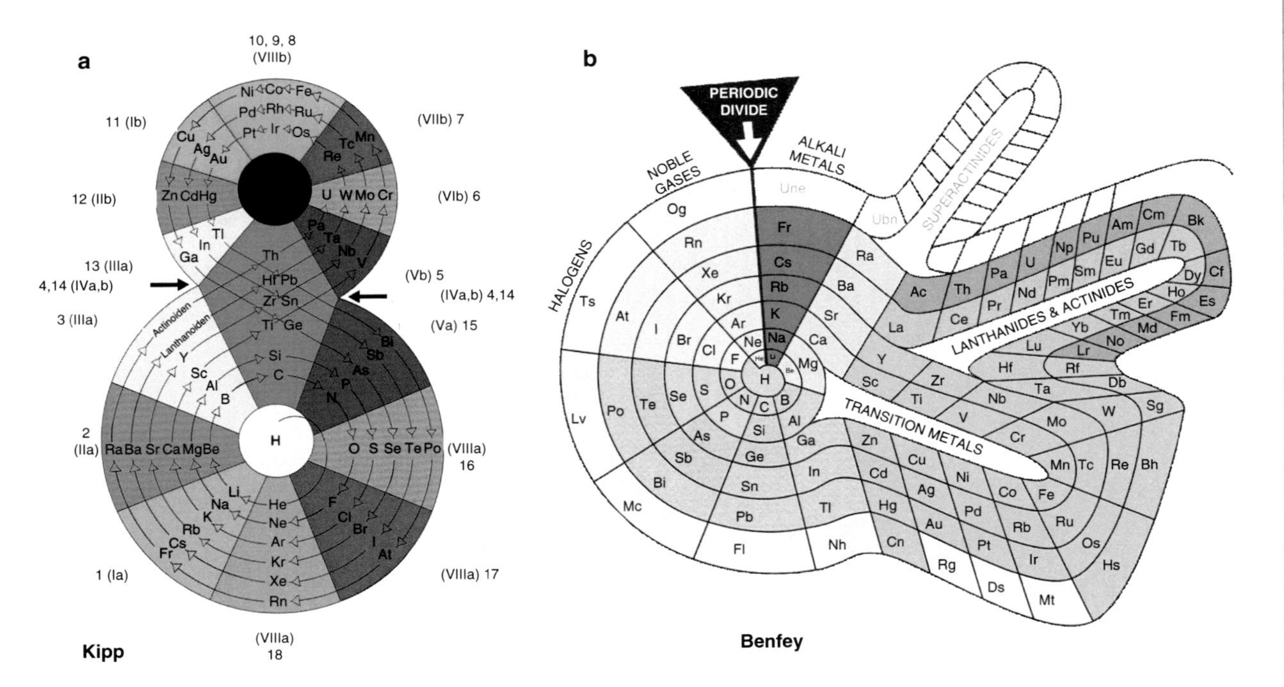

Fig. 6.3 (**a**) Spiral period system according to Kipp (1942, revised), (**b**) Loop period system according to Benfey. (Source: By Mardeg in English Wikipedia, CC BY-SA 3.0, https://commons.wikimedia.org/w/index.php?curid=6464611); see Sect. 6.3 for details

Period	Group I a	Group I b	Group II a	Group II b	Group III a	Group III b	Group IV a	Group IV b	Group V a	Group V b	Group VI a	Group VI b	Group VII a	Group VII b	Group VIII			Group 0
I	1 **H** 1,0081																	2 **He** 4,003
II	3 **Li** 6,940		4 **Be** 9,02			5 **B** 10,82		6 **C** 12,01		7 **N** 14,008		8 **O** 16,000		9 **F** 19,00				10 **Ne** 20,183
III	11 **Na** 22,997		12 **Mg** 24,32			13 **Al** 26,97		14 **Si** 28,06		15 **P** 30,98		16 **S** 32,06		17 **Cl** 35,457				18 **Ar** 39,944
IV	19 **K** 39,096		20 **Ca** 40,08		21 **Sc** 45,10		22 **Ti** 47,90		23 **V** 50,95		24 **Cr** 52,01		25 **Mn** 54,93		26 **Fe** 55,85	27 **Co** 58,94	28 **Ni** 58,69	
		29 **Cu** 63,57		30 **Zn** 65,38		31 **Ga** 69,72		32 **Ge** 72,60		33 **As** 74,91		34 **Se** 78,96		35 **Br** 79,916				36 **Kr** 83,7
V	37 **Rb** 85,48		38 **Sr** 87,63		39 **Y** 88,92		40 **Zr** 91,22		41 **Nb** 92,91		42 **Mo** 95,95		43 **Tc** (98)		44 **Ru** 101,7	45 **Rh** 102,91	46 **Pd** 106,7	
		47 **Ag** 107,880		48 **Cd** 112,41		49 **In** 114,76		50 **Sn** 118,70		51 **Sb** 121,76		52 **Te** 127,61		53 **I** 126,92				54 **Xe** 131,3
VI	55 **Cs** 132,91		56 **Ba** 137,36		57-71 Seltene Erden*		72 **Hf** 178,6		73 **Ta** 180,88		74 **W** 183,92		75 **Re** 186,31		76 **Os** 190,2	77 **Ir** 193,1	78 **Pt** 195,23	
		79 **Au** 197,2		80 **Hg** 200,61		81 **Tl** 204,39		82 **Pb** 209,00		83 **Bi** 209,00		84 **Po** (210)		85 **At** (210)				86 **Rn** 222
VII	87 **Fr** (223)		88 **Ra** 226,05		89 **Ac** (227)		90 **Th** 232,12		91 **Pa** 231		92 **U** 238,07							

* Rare earths:	57 **La**	58 **Ce**	59 **Pr**	60 **Nd**	61 **??**	62 **Sm**	63 **Eu**	64 **Gd**	65 **Tb**	66 **Dy**	67 **Ho**	68 **Er**	69 **Tm**	70 **Yb**	71 **Lu**
	138,92	140,13	140,92	144,27	–	150,43	152,0	156,9	159,2	162,46	163,5	167,2	169,4	173,04	174,99

Group designations: Ia alkali metals, Ib copper group, IIa alkaline earth metals, IIb without special name, IIIa rare earths, IIIb earth metals, IVa earth acid formers Gr. IV, IVb carbon group, Va Earth acidifiers Gr. V, Vb Nitrogen group, VIa Acidifiers Gr. VI, VIb Oxygen group, VIIa Acidifiers Gr. VII, VIIb Halogens, VIII Iron and platinum metals, 0 Noble gases

Fig. 6.4 Short period system (ca. 1950, new drawing), for explanations see Sect. 6.4

6.5 Periodic Table According to A. V. Antropov

A mixed form between the long period system, which is in use today, and the short period system goes back to A. V. Antropow (1926). Here the periods are arranged in columns of different widths, but strictly according to atomic number (Fig. 6.5). In addition, the noble gases are present twice: As the eighth and as the zeroth group. The bottom row shows the current IUPAC numbering.

6.6 Split Long-Period System

The usual representation of the periodic table today is the so-called long form (compare Fig. 5.1). In this variant, the secondary groups ("d elements") are placed between the main groups and the rare earths (lanthanides and actinides; "f elements") are arranged in two lines below them.

For didactic reasons, a split, almost pyramidal representation according to Fig. 6.6 is sometimes used. Here the three blocks (main groups, subgroups and 'rare earths') are clearly separated from each other and the places where the 'insertions' occur are clearly marked.

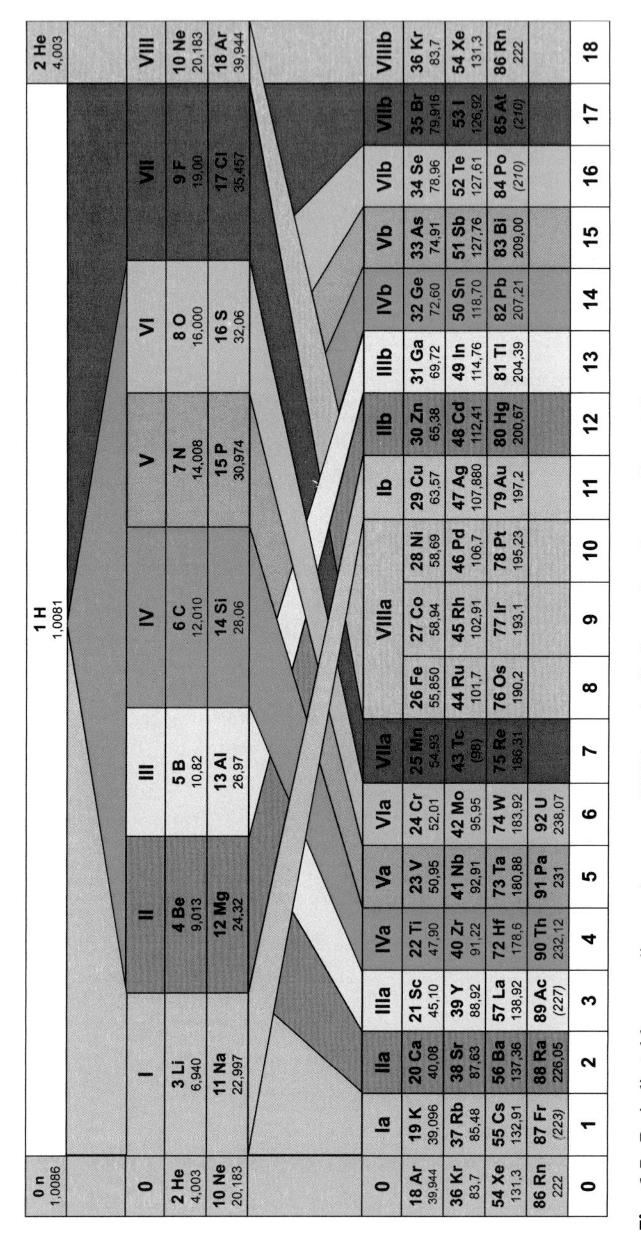

Fig. 6.5 Periodic table according to Antropov (1926, revised), for explanation see Sect. 6.5

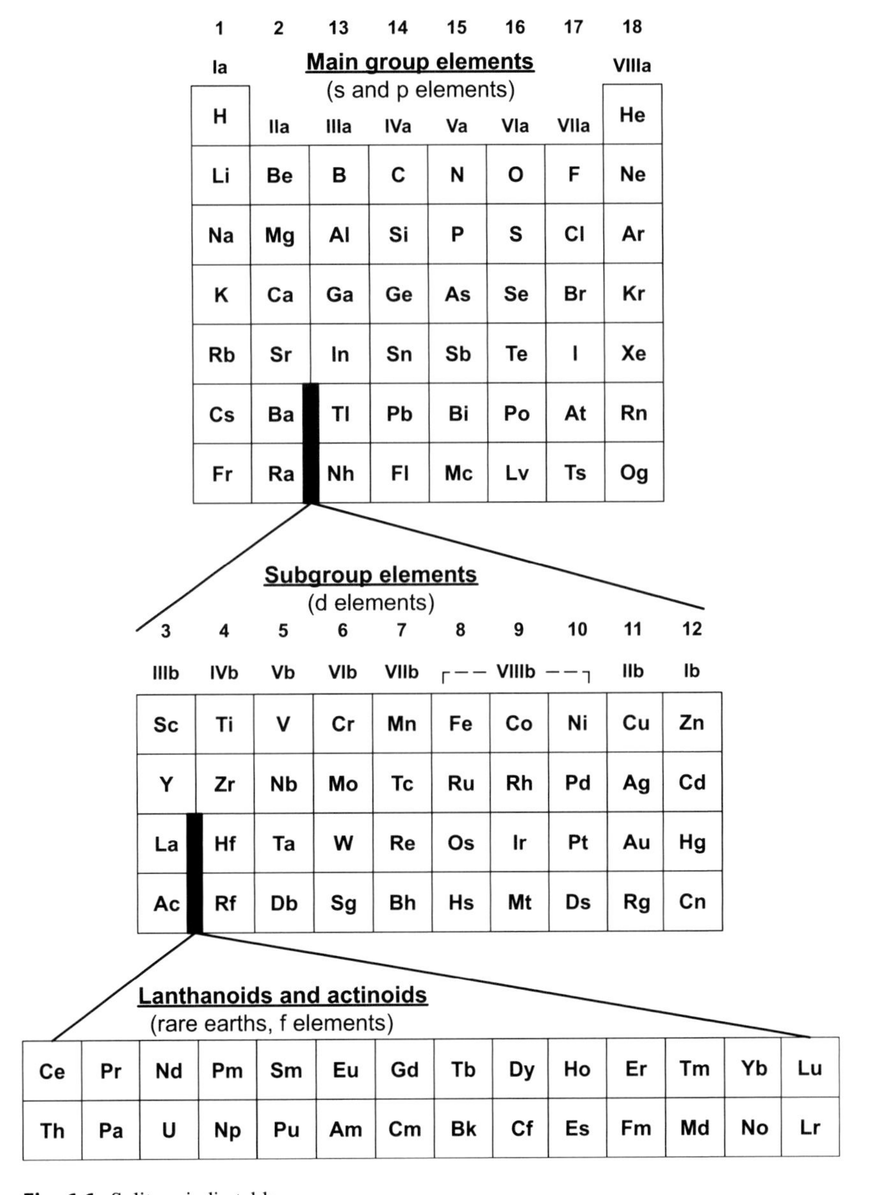

Fig. 6.6 Split periodic table

What You Learned from This *essential*

- Historical sequence of the discovery of the periodic table
- Explanation of periodicity within the PTE
- Trends of the periodicities within the periodic table
- Examples of the transformation of the periodic table and alternative ways of representation

T. Schmiermund, *The discovery of the periodic table of the chemical elements*, essentials, https://doi.org/10.1007/978-3-658-36448-9

References

Binder HH (1999) Lexikon der chemischen Elemente. S. Hirzel, Stuttgart

Binnewies M, Jäckel M, Willner H, Rayner-Canham G (2004) Allgemeine und Anorganische Chemie. Spektrum, Heidelberg

Christen HR, Meyer G (1997) Grundlagen der Allgemeinen und Anorganischen Chemie. Salle + Sauerländer, Frankfurt a. M.

Cotton FA, Wilkinson G, Gaus PL (1990) Grundlagen der Anorganischen Chemie. Wiley-VCH, Weinheim (English edition: Basic inorganic chemistry, 2nd ed. Wiley, 1987)

Dickerson RE, Gray HB, Haight GP (1978) Prinzipien der Chemie. De Gruyter, Berlin (English Edition: Chemical principles, 2nd ed. W. A. Benjamin, 1974)

Döbereiner JW (1829) Versuch zu einer Gruppierung der elementaren Stoffe nach ihrer Analogie. Ann Phys 15:301–307. (Poggendorff JC (Hrsg))

Falbe J, Regitz M (Hrsg) (1995) Römpp Chemie Lexikon, 9. Aufl. Georg Thieme, Stuttgart

Felixberger JK (2017) Chemie für Einsteiger. Springer, Heidelberg

Fluck E, Heumann KG (2012) Periodensystem der Elemente: physikalische Eigenschaften, 5. Aufl. Wiley-VCH, Weinheim

GDCh Gesellschaft Deutscher Chemiker e. V. (Hrsg) (2019) Elemente – 150 Jahre Periodensystem. Sonderdruck für Spektrum der Wissenschaft, Frankfurt a. M. und Heidelberg. https://www.gdch.de/service-information/jahr-des-pse.html

Greenwood NN, Earnshaw A (1988) Chemie der Elemente. Wiley-VCH, Weinheim (English edition: Chemistry of the elements. Pergamon Press, Oxford, 1984)

Hardt H-D (1987) Die periodischen Eigenschaften der chemischen Elemente. Thieme, Stuttgart

Hollemann AF, Wiberg E (1985) Lehrbuch der anorganischen Chemie, 91.–100. Aufl. De Gruyter, Berlin

Langhammer G (1949) Das Periodensystem. Volk und Wissen, Leipzig

Latscha HP, Klein HA (1996) Anorganische Chemie, 7. Aufl. Springer, Heidelberg

Latscha HP, Mutz M (2011) Chemie der Elemente, 1. Aufl. Springer, Heidelberg

Leach M (1996–2019) Internet database of periodic tables. https://www.meta-synthesis.com/webbook/35_pt/pt_database.php. Last accessed 23 Oct 2019

Mendelejeff D (1869a) Ueber die Beziehungen der Eigenschaften zu den Atomgewichten der Elemente, abgedruckt in: Seubert K (Hrsg) (1895) Das natürliche System der chemischen

T. Schmiermund, *The discovery of the periodic table of the chemical elements*, essentials, https://doi.org/10.1007/978-3-658-36448-9

Elemente, Reprint der Orig.-Ausg. (1996), Harri Deutsch, Frankfurt a. M. (Ostwalds Klassiker der exakten Wissenschaften, Bd 68)
Mendelejeff D (1869b) Die Beziehungen zwischen den Eigenschaften der Elemente und ihren Atomgewichten, abgedruckt in: Seubert K (Hrsg) (1895) Das natürliche System der chemischen Elemente, Reprint der Orig.-Ausg. (1996), Harri Deutsch, Frankfurt a. M. (Ostwalds Klassiker der exakten Wissenschaften, Bd 68)
Mendelejeff D (1871) Die periodische Gesetzmäßigkeit der chemischen Elemente, abgedruckt in: Seubert K (Hrsg) (1895) Das natürliche System der chemischen Elemente, Reprint der Orig.Ausg. (1996), Harri Deutsch, Frankfurt a. M. (Ostwalds Klassiker der exakten Wissenschaften, Bd 68)
Meyer L (1864) Natur der Atome: Gründe gegen ihre Einfachheit, abgedruckt in: Seubert K (Hrsg) (1895) Das natürliche System der chemischen Elemente, Reprint der Orig.-Ausg. (1996), Harri Deutsch, Frankfurt a. M. (Ostwalds Klassiker der exakten Wissenschaften, Bd 68)
Meyer L (1870) Die Natur der chemischen Elemente als Function ihrer Atomgewichte, abgedruckt in: Seubert K (Hrsg) (1895) Das natürliche System der chemischen Elemente, Reprint der Orig.-Ausg. (1996), Harri Deutsch, Frankfurt a. M. (Ostwalds Klassiker der exakten Wissenschaften, Bd 68)
Mortimer CE (1996) Chemie, 6. Aufl. Georg Thieme, Stuttgart (English edition: Chemistry. Wadsworth Publishing, Belmont, 1996)
Pötsch WR, Fischer A, Müller W (1988) Lexikon bedeutender Chemiker. VEB Bibliographisches Institut, Leipzig
Riedel E (1999) Anorganische Chemie. De Gruyter, Berlin
Scerri E (2007) The periodic table. Its story and its significance. Oxford University Press, Oxford
Schmiermund T (2019) Das Chemiewissen für die Feuerwehr. Springer, Heidelberg
Van Spronsen JW (1965) The periodic system of chemical elements. A history of the first hundred years. Elsevier, New York
Wawra E, Dolznig H, Müllner E (2010) Chemie erleben, 2. Aufl. Facultas, Wien
Welsch N, Schwab J, Liebmann CC (2013) Materie – Erde, Wasser, Luft und Feuer. Springer, Heidelberg
Weyer J (2018) Geschichte der Chemie, Bd 2. Springer, Heidelberg
Wußing H-L (1983) Geschichte der Naturwissenschaften. Druckerei Fortschritt, Erfurt
Wußing H-L, Diertich H, Purkert W, Tutzke D (eds) (1992) Fachlexikon Forscher und Erfinder. Harri Deutsch, Frankfurt a. M

Printed in the United States
by Baker & Taylor Publisher Services